TALKING
with
PLANETS

NIKOLA TESLA

POCKETSIZE PUBLISHING

Nikola Tesla

Editor's Note. — *Mr. Nikola Tesla has accomplished some marvelous results in electrical discoveries. Now, with the dawn of the new century, he announces an achievement which will amaze the entire universe, and which eclipses the wildest dream of the most visionary scientist. He has received communication, he asserts, from out the great void of space; a call from the inhabitants of Mars, or Venus, or some other sister planet! And, furthermore, noted scientists, like Sir Norman Lockyer are disposed to agree with Mr. Tesla in his startling deductions.*

Mr. Tesla has not only discovered many important principles, but most of his inventions are in practical use; notably in the harnessing of the Titanic forces of Niagara Falls, and the discovery of a new light by means of a vacuum tube. He has, he declares, solved the problem of telegraphing without wires or artificial conductors of any sort, using the earth as his medium. By means of this principle he expects to be able to send messages under the ocean, and to any distance on the earth's surface. Interplanetary communication has interested him for years, and he sees no reason why we should not soon be within talking distance of Mars or of all worlds in the solar system that may be tenanted by intelligent beings.

At the request of "Collier's Weekly," Mr. Tesla presents herewith a frank statement of what he expects to accomplish and how he hopes to establish communication with the planets.

The idea of communicating with the inhabitants of other worlds is an old one. But for ages it has been regarded merely as a poet's dream, forever unrealizable. And yet, with the invention and perfection of the telescope and the ever-widening knowledge of the heavens, its hold upon our imagination has been increased, and the scientific achievements during the latter part of the nineteenth century, together with the development of the tendency toward the nature ideal of Goethe, have intensified it to such a degree that it seems as if it were destined to become the dominating idea of the century that has just begun. The desire to know something of our neighbors in the immense depths of space does not spring from idle curiosity nor from thirst for knowledge, but from a deeper cause, and it is a feeling firmly rooted in the heart of every human being capable of thinking at all.

Whence, then, does it come? Who knows? Who can assign limits to the subtlety of nature's influences? Perhaps, if we could clearly perceive all the intricate mechanism of the glorious spectacle that is continually unfolding before us, and could, also, trace this desire to its distant origin, we might find it in the sorrowful vibrations of the earth which began when it parted from its celestial parent.

*Photographic view of the essential parts of the
electrical oscillator used in Mr. Tesla's experiments.*

But in this age of reason it is not astonishing to find
persons who scoff at the very thought of effecting
communication with a planet. First of all, the argument
is made that there is only a small probability of other
planets being inhabited at all. This argument has
never appealed to me. In the solar system, there seem
to be only two planets — Venus and Mars — capable
of sustaining life such as ours: but this does not mean
that there might not be on all of them some other
forms of life. Chemical processes may be maintained
without the aid of oxygen, and it is still a question
whether chemical processes are absolutely necessary to
the sustenance of organised beings. My idea is that the

development of life must lead to forms of existence that will be possible without nourishment and which will not be shackled by consequent limitations. Why should a living being not be able to obtain all the energy it needs for the performance of its life-functions from the environment, instead of through consumption of food, and transforming, by a complicated process, the energy of chemical combinations into life-sustaining energy?

If there were such beings on one of the planets we should know next to nothing about them. Nor is it necessary to go so far in our assumptions, for we can readily conceive that, in the same degree as the atmosphere diminishes in density, moisture disappears and the planet freezes up, organic life might also undergo corresponding modifications, leading finally to forms which, according to our present ideas of life, are impossible. I will readily admit, of course, that if there should be a sudden catastrophe of any kind all life processes might be arrested; but if the change, no matter how great, should be gradual, and occupied ages, so that the ultimate results could be intelligently foreseen, I cannot but think that reasoning beings would still find means of existence. They would adapt themselves to their constantly changing environment. So I think it quite possible that in a frozen planet, such as our moon is supposed to be, intelligent beings may still dwell, in its interior, if not on its surface.

SIGNALING AT 100,000,0%00 MILES!

*Supplying electrical energy through a
single wire without return.*

Then it is contended that it is beyond human power and ingenuity to convey signals to the almost inconceivable distances of fifty million or one hundred million miles. This might have been a valid argument formerly. It is not so now. Most of those who are enthusiastic upon the subject of interplanetary communication have reposed their faith in the light-ray as the best possible medium of such communication. True, waves of light, owing to their immense rapidity of succession, can penetrate space more readily than waves less rapid, but a simple consideration will show that by their means an exchange of signals between this earth and its companions in the solar system is, at least now, impossible. By way of illustration, let us suppose that a square mile of the earth's surface — the smallest area that might possibly be within reach of the best telescopic vision of other worlds — were covered with incandescent lamps, packed closely together, so as to form, when illuminated, a continuous sheet of light. It would require not less than one hundred million horse-power to light this area of lamps, and this is many times the amount of motive power now in the service of man throughout the world.

But with the novel means, proposed by myself, I can readily demonstrate that, with an expenditure not exceeding two thousand horse-power, signals can be transmitted to a planet such as Mars with as much exactness and certitude as we now send messages by wire from New York to Philadelphia. These means are the result of long-continued experiment and gradual improvement.

Some ten years ago, I recognised the fact that to convey electric currents to a distance it was not at all necessary to employ a return wire, but that any amount of energy might be transmitted by using a single wire. I illustrated this principle by numerous experiments, which, at that time, excited considerable attention among scientific men.

This being practically demonstrated, my next step was to use the earth itself as the medium for conducting the currents, thus dispensing with wires and all other artificial conductors. So I was led to the development of a system of energy transmission and of telegraphy without the use of wires, which I described in 1893. The difficulties I encountered at first in the transmission of currents through the earth were very great. At that time I had at hand only ordinary apparatus, which I found to be ineffective, and I concentrated my attention immediately upon perfecting machines for this special purpose. This work consumed a number of years, but I finally vanquished all difficulties and succeeded in producing a machine which, to explain its operation in plain language, resembled a pump in its action, drawing electricity from the earth and driving it back into the same at an enormous rate, thus creating ripples or disturbances which, spreading through the earth as through a wire, could be detected at great distances by carefully attuned receiving circuits. In this manner I was able to transmit to a distance, not only feeble effects for the purposes of signalling, but considerable amounts of energy, and later discoveries I made convinced me that I shall ultimately succeed in conveying power without wires, for industrial purposes, with high economy, and to any distance, however great.

EXPERIMENTS IN COLORADO

An electrical oscillator delivering energy at a rate of 75,000 horse-power.

To develop these inventions further, I went to Colorado in 1899, where I continued my investigations along these and other lines, one of which, in particular, I now consider of even greater importance than the transmission of power without wires. I constructed a laboratory in the neighborhood of Pike's Peak. The conditions in the pure air of the Colorado Mountains proved extremely favorable for my experiments, and the results were most gratifying to me. I found that I could not only accomplish more work, physically and mentally, than I could in New York, but that electrical effects and changes were more readily and distinctly perceived. A few years ago it was virtually impossible to produce electrical sparks twenty or thirty foot long; but I produced some more than one hundred feet in length, and this without difficulty. The rates of electrical movement involved in strong induction apparatus had measured but a few hundred horse-power, and I produced electrical movements of rates of one hundred and ten thousand horse-power. Prior to this, only insignificant electrical pressures were obtained, while I have reached fifty million volts.

The accompanying illustrations, with their descriptive titles, taken from an article I wrote for the "Century Magazine," may serve to convey an idea of the results I obtained in the directions indicated.

Many persons in my own profession have wondered at them and have asked what I am trying to do. But the time is not far away now when the practical results of my labors will be placed before the world and their influence felt everywhere. One of the immediate consequences will be the transmission of messages without wires, over sea or land, to an immense distance. I have already demonstrated, by crucial tests, the practicability of signalling by my system from one to any other point of the globe, no matter how remote, and I shall soon convert the disbelievers.

I have every reason for congratulating myself that throughout these experiments, many of which were exceedingly delicate and hazardous, neither myself nor any of my assistants received any injury. When working with these powerful electrical oscillations the most extraordinary phenomena take place at times. Owing to some interference of the oscillations, veritable balls of fire are apt to leap out to a great distance, and if any one were within or near their paths, he would be instantly destroyed. A machine such as I have used could easily kill, in an instant, three hundred thousand persons. I observed that the strain upon my assistants was telling, and some of them could not endure the extreme tension of the nerves. But these perils are now entirely overcome, and the operation of such apparatus, however powerful, involves no risk whatever.

As I was improving my machines for the production of intense electrical actions, I was also perfecting the means for observing feeble effects. One of the most interesting results, and also one of great practical importance, was the development of certain contrivances for indicating at a distance of many hundred miles an approaching storm, its direction, speed and distance travelled.

These appliances are likely to be valuable in future meteorological observations and surveying, and will lend themselves particularly to many naval uses.

It was in carrying on this work that for the first time I discovered those mysterious effects which have elicited such unusual interest. I had perfected the apparatus referred to so far that from my laboratory in the Colorado mountains I could feel the pulse of the globe, as it were, noting every electrical change that occurred within a radius of eleven hundred miles.

TERRIFIED BY SUCCESS

Transmitting electrical energy
through the earth without wire.

I can never forget the first sensations I experienced when it dawned upon me that I had observed something possibly of incalculable consequences to mankind. I felt as though I were present at the birth of a new knowledge or the revelation of a great truth. Even now, at times, I can vividly recall the incident, and see my apparatus as though it were actually before me. My first observations positively terrified me, as there was present in them something mysterious, not to say supernatural, and I was alone in my laboratory at night; but at that time the idea of these disturbances being intelligently controlled signals did not yet present itself to me.

The changes I noted were taking place periodically, and with such a clear suggestion of number and order that they were not traceable to any cause then known to me. I was familiar, of course, with such electrical disturbances as are produced by the sun, Aurora Borealis and earth currents, and I was as sure as I could be of any fact that these variations were due to none of these causes. The nature of my experiments precluded the possibility of the changes being produced by atmospheric disturbances, as has been rashly asserted by some. It was some time afterward when the thought flashed upon my mind that the disturbances I had observed might be due to an intelligent control. Although I could not decipher their meaning, it was impossible for me to think of them as having been entirely accidental. The feeling is constantly growing on me that I had been the first to hear the greeting of one planet to another. A purpose was behind these electrical signals; and it was with this conviction that I announced to the Red Cross Society, when it asked me to indicate one of the great possible achievements of the next hundred years, that it would probably be the confirmation and interpretation of this planetary challenge to us.

Since my return to New York more urgent work has consumed all my attention; but I have never ceased to think of those experiences and of the observations made in Colorado. I am constantly endeavoring to improve and perfect my apparatus, and just as soon as practicable I shall again take up the thread of my investigations at the point where I have been forced to lay it down for a time.

COMMUNICATING WITH THE MARTIANS

At the present stage of progress, there would be no insurmountable obstacle in constructing a machine capable of conveying a message to Mars, nor would there be any great difficulty in recording signals transmitted to us by the inhabitants of that planet, if they be skilled electricians. Communication once established, even in the simplest way, as by a mere interchange of numbers, the progress toward more intelligible communication would be rapid. Absolute certitude as to the receipt and interchange of messages would be reached as soon as we could respond with the number "four," say, in reply to the signal "one, two, three." The Martians, or the inhabitants of whatever planet had signalled to us, would understand at once that we had caught their message across the gulf of space and had sent back a response. To convey a knowledge of form by such means is, while very difficult, not impossible, and I have already found a way of doing it.

Experiment to illustrate the capacity of the oscillator for producing electrical explosions of great power.

What a tremendous stir this would make in the world! How soon will it come? For that it will some time be accomplished must be clear to every thoughtful being.

Something, at least, science has gained. But I hope that it will also be demonstrated soon that in my experiments in the West I was not merely beholding a vision, but had caught sight of a great and profound truth.

www.ingramcontent.com/pod-product-compliance
Lightning Source LLC
Chambersburg PA
CBHW050355210326
41520CB00020B/6322